I0025450

Pollination Parade

Brooklyn Richards

Copyright © 2024 by Brooklyn Richards
All rights reserved.
ISBN 979-8-218-55243-5

For my honey,
Mitch.

A is for **Animal kingdom**. All animals, from giant squids to tiny fruit flies, belong to the animal kingdom. Some animals are pollinators. That means they help flowers grow by moving pollen around. Bees, birds, and butterflies are just a few! Flowers say, "thank you" by giving pollinators a special treat called nectar.

B is for **Bat**. Bats are incredible critters. They're the only mammals that can fly, and they have a special power to see in the dark using sound and echoes! This is called echolocation. When bats visit plants like agave and cacti, their fuzzy faces get covered in pollen.

FLAVORS

1. Chocolate
2. Caramel swirl
3. Raspberry
4. Pistachio
5. Cookie Dough
6. Strawberry
7. Vanilla
8. Brownie
9. Mint chip
10. Bubblegum
11. Birthday cake
12. Chocolate+Cherry

C is for **Courtship**. Love is in the air! There are lots of ways pollinators try to attract a mate. This hummingbird is putting on a special show to impress his crush. He flies up high and dives back down while flashing his fancy feathers.

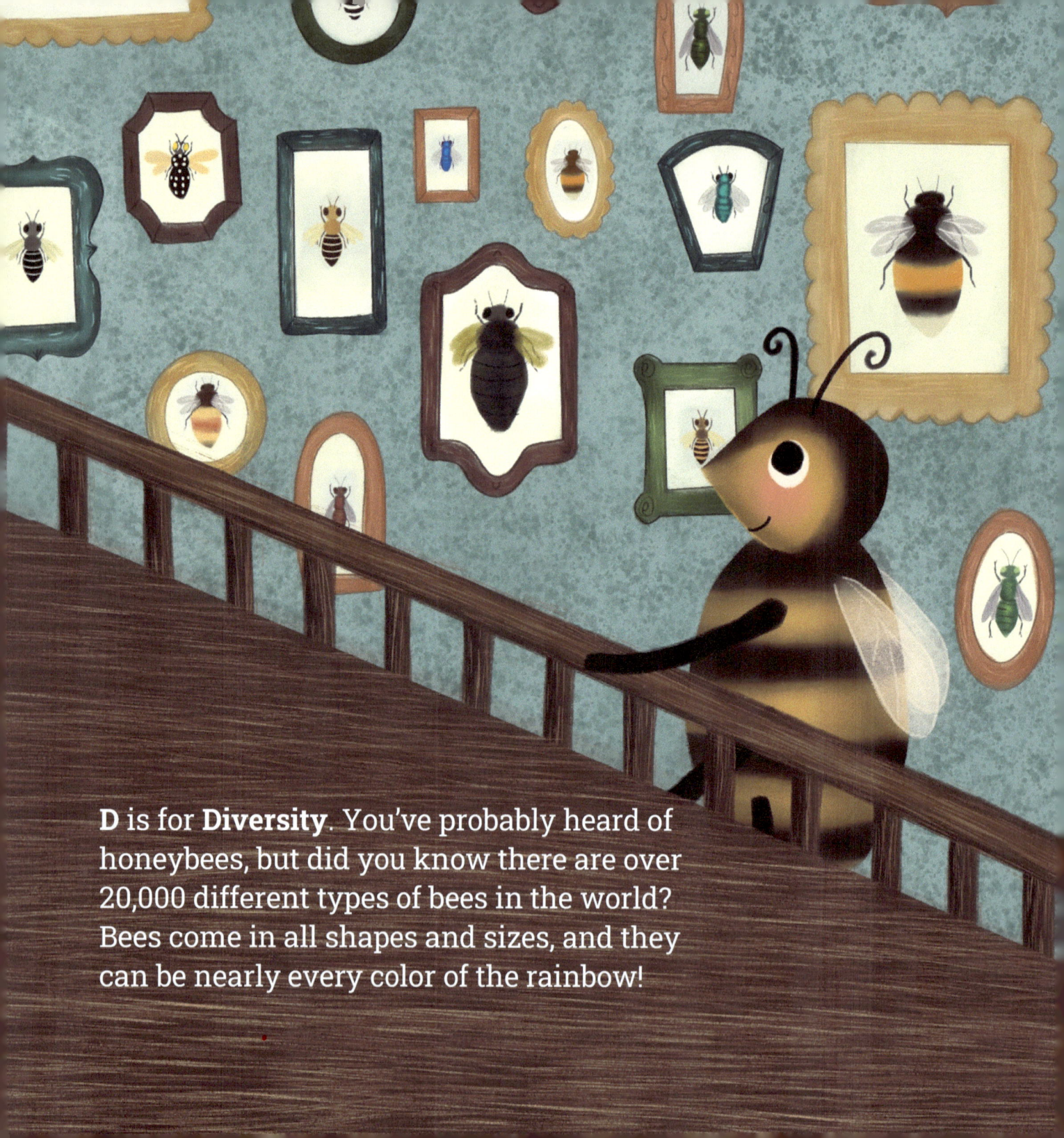

D is for **Diversity**. You've probably heard of honeybees, but did you know there are over 20,000 different types of bees in the world? Bees come in all shapes and sizes, and they can be nearly every color of the rainbow!

E is for **Ecosystem**. An ecosystem connects plants, animals, and the places they call home. Everything in an ecosystem has a special job, and pollinators have one of the coolest jobs of all! With the help of pollinators, plants can grow seeds and yummy fruit.

F is for **Fly**. Buzz, buzz—what's that sound?
Move over, bees! Flies like flowers, too! Flies
have fuzzy bodies perfect for picking up pollen.

G is for **Garden**. You can be a pollinator hero by planting their favorite flowers. Different flowers help different types of pollinators. Some pollinators love many types of plants, while others are quite picky. Plants that are native to where you live are a great choice for your pollinator garden.

H is for **Hotel**. A bee hotel, that is! Many bees, like mason bees and leafcutter bees, build their nests inside of stems or holes in wood. You can give bees a safe, cozy place to lay their eggs by putting a bee hotel in your garden.

On the chalkboard:

FLOWER PARTS:

PETAL

STAMEN

PISTIL

SEPAL

STEM

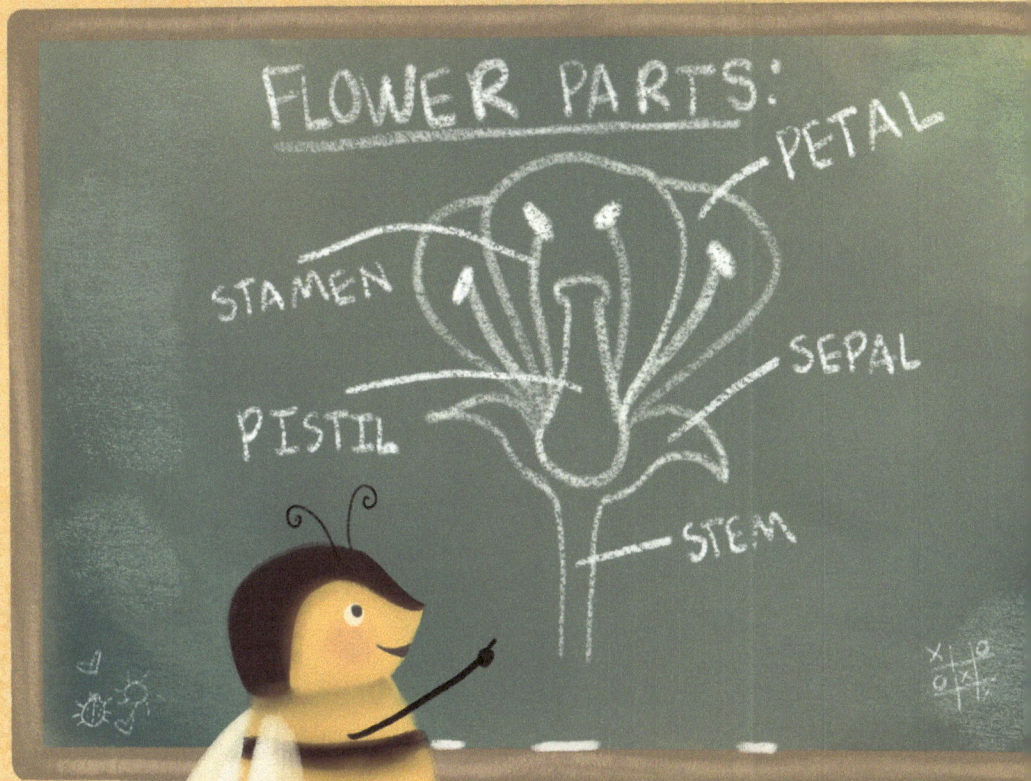

I is for **Inside the flower**. What do the parts inside the flower do? Let's take a look! The stamen makes pollen, while the pistil is where fruit and seeds grow.

J is for **Journey**. Choo-choo! All aboard the monarch migration! These butterflies are on an adventure. Each year, they fly south to find warmer places in the winter and the next generation returns home in the spring.

K is for **Knowing where to go**. Pollinators have lots to keep track of, from finding flowers to remembering the way back home! Bees pay close attention to what's around them and find their way using the position of the sun.

L is for **Learning who's who**. Lots of pollinators look alike. Sometimes even experts have to take a closer look! Can you spot differences between this bee, fly, and wasp? Noticing details like body shape, hair, wings, and antennae can help you get to know the pollinators in your neighborhood.

M is for **Mimic**. Some pollinators wear clever costumes to disguise as someone scarier. Stripes like a bee or wings like a monarch tell predators, "Stay away!"

HALLOWE[EN]
PARTY

OCTOBER 3[1]
Costume conte[st]

FACE
PAINT

N is for **Nighttime**. When the sun goes down, nocturnal pollinators get their groove on! Bats and moths take the night shift, visiting flowers that bloom after dark.

O is for **Offspring**. Pollinators work hard to build the perfect nest for their babies. Leafcutter bees cut out tiny pieces of leaves to line their nests, while hummingbirds weave cozy homes using soft things like spiderwebs and moss.

Fresh
cut
leaves

suzy's
Spider webs

P is for **Protection**. Some pollinators have built-in defenses to keep them safe. Bees, for example, have stingers to warn off predators who get too close. But don't worry—bees are usually gentle, and they won't sting us if we stay calm and don't bother them.

Q is for **Queen**. Honeybees and bumblebees have a queen who leads the hive. Her job is to lay eggs, while worker bees care for her, gather pollen, and protect the hive. But not all bees live in hives. Most are solitary and work alone!

R is for **Robbing**. Sneaky visitors, called "nectar robbers," steal nectar without helping to pollinate! They bite a hole in the side of the flower to sip nectar without touching the pollen.

Must be this tall to ride

caramel apples

Crysalis Candy Co.

entrance

S is for **Size**. One size does not fit all when it comes to pollination! Tiny pollinators are great at fitting into small flowers, while those with bigger bodies can carry lots of pollen. Big or small, every pollinator has something special about them.

T is for **Treat**. Nectar is the sweet reward that flowers provide to their helpful visitors. Nectar gives pollinators the energy they need to go, go, go!

U is for **Underground**. Did you know that most types of bees build their homes underground? These bees burrow into the soil to lay their eggs and store food. Their underground nest keeps them safe and snug.

V is for **Vibration**. You've probably heard bees buzzing around in the garden, but some have a secret buzzing superpower! These bees are called buzz pollinators, and they use a special type of buzz to shake pollen off flowers.

W is for **Waggle dance**. Honeybees have big hives with lots of mouths to feed, and it takes teamwork to get the job done! The waggle dance is how honeybees communicate about finding flowers. When a bee does the waggle dance, it will run, wiggle, and spin around to show other bees where to go.

X is for **eXcavation**. Building a home is hard work! Ground-nesting bees dig cozy tunnels beneath the soil, while carpenter bees use their strong jaws to chew holes in wood.

STOP

Y is for **Yum**. Most of our favorite foods wouldn't be here without the help of pollinators! From delicious berries to tasty chocolate, pollinators help keep our bellies full and our planet healthy.

Z is for **Zzz....** these pollinators sure are sleepy after a long day! They find a cozy place to rest before waking up to do it all over again. Sweet dreams, little pollinators!

www.ingramcontent.com/pod-product-compliance
Lightning Source LLC
Chambersburg PA
CBHW060853270326
41934CB00002B/125